Images
of
Barry Scrapyard

By
Roger Hardingham

First Published 1999

ISBN 0 946184 86 0

Front cover: The scene at Barry in August 1980, when the cutting up of
locomotives was at the fore once again. The 'Merchant Navy' Class on
the right, No. 35025 *Brocklebank Line*, was going to be the next one if the
reprieve had not come at the eleventh hour.

All photographs by Roger Hardingham, unless otherwise stated.

Published By
Waterfront
A Division of Kingfisher Productions
The Dalesmade Centre, Watershed Mill, Settle,
North Yorkshire BD24 9LR
Printed by Ian Allan Printing, London

Introduction

The very existence of preserved railways in Britain today is solely due to a miracle occurring in the town of Barry in South Glamorgan.

It all began in 1959 and lasted for exactly 30 years, until 1989. Up until the late 1950s Barry was content with its life as an important South Wales port for the export of huge quantities of coal and latterly with the import of such foodstuffs as grain and bananas. But working within the dock area was the firm of Woodham Brothers, well known locally as general merchants and for their recycling of ship's materials. However, in the late 1950s the scrap business in Britain was to see a potential gold mine develop as British Railways carried out their restructuring plan.

The far reaching plan was to see a modernised railway network in Britain, eliminating almost 16,000 steam locomotives and hundreds of thousands of outdated freight vehicles. The steam era was due to be eradicated by 1968.

The little firm of Woodham Brothers could see the potential and were quick to 'get onto the gravy train', as Dai Woodham later put it. At the invitation of Swindon Works, Dai and other potential scrap merchants went to the capital of the Great Western to see and learn how they cut up locomotives. All the Works in the different regions, were hitherto responsible for this breaking up. But as the Modernisation Plan took hold, these railway facilities were clearly unable to keep pace with the huge influx of redundant locomotives. So, on 22nd February 1959, the Western Region was the first to sell a batch of engines to a private contractor and Woodhams were the first to receive this new business.

The first batch of five were dispatched from the dump at Swindon and arrived on 25th March 1959; all were broken up by the end of the summer.

In the next few years purchases of condemned locomotives gathered pace, although it must be said that Woodhams was never the largest of the scrap companies; in total they only received some 297, of which they actually scrapped 84. (Cashmore's at nearby Newport also began in 1959 by receiving their first four ex-GWR 2-6-2Ts on 9th April, but in total they cut up over 1,000 locomotives.) It was the

Four rows of locomotives as far as the eye can see in the lower, West Pond, site at Barry in the 1980s. The rain often provided a better photographic opportunity of the locomotives. Photo: Glynn Hague

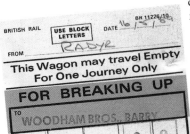

gaining of lucrative wagon contracts though that really kept Woodhams' cutting torches busy. These contracts would supply a steady stream of work from the mid-1960s and as history now tells us, this left the locomotives as, 'work in hand', for many of the remaining years.

These factors and the end of steam itself in Britain in August 1968, came together and set the course for a 30-year phenomenon, never likely to be repeated. As steam vanished from our railways, supporters of this motive power started to realise the loss and as the 1960s progressed, this loss became increasingly evident. Sterling work had been accomplished in the late 1950s and early 1960s to save some shining examples of steam, but the hard reality was that most had disappeared before it was too late to do anything about it. This is where Barry stepped in.

As locomotives were still arriving at Barry from the London Midland Region, one ex-LMS locomotive was departing; No. 43924 was the first to be bought from Woodhams for preservation and departed for the Keighley & Worth Valley Railway in September 1968.

This was a significant occasion and was quite quickly followed, in January 1969, by the removal of Southern 'U' Class No. 31618. It had been at Barry just four and a half years. These two purchases and another that year in the form of ex-GWR No. 5322, set the stage for the following twenty years.

People started to realise that, yes you could buy an ex-S&D 2-8-0 in Britain still, or a Southern 'Q', or maybe even a 'Hall' or 'Jubilee'. The railway press became full of advertisements appealing for funds to buy a particular locomotive and fortunately most societies were successful, even if it meant trailing around the country selling books, pens and badges to raise the all important cash. The crucial thing was, if you did raise say £2,500 or perhaps £4,000, you were quite capable of buying a main line tender locomotive.

The early 1970s saw a boom in the buying and moving of locomotives from Woodhams yard; 10 left in 1972; 18 in 1973; 19 in 1974 and 14 in 1975. During this period two locomotives were actually cut up. These were No. 76080 in April 1972 and No. 3817 in March 1973. There was no real pattern developing with departures, but clearly some locomotives were bought but not moved straight away, some stayed in the yard for several years whilst the owner raised the haulage costs. In the early 1970s an odd form of levy was slapped on to a locomotive's cost. This was a result of BR stipulating to Woodhams that the locomotives were for scrap only and not for resale! This was fortunately dropped soon afterwards, but then the dreaded VAT came into force on 1st April 1973, adding 10% onto your bill. If this wasn't enough, the price of scrap metal was deregulated, which together with the depression in scrap metal requirements and general inflation

An ex-GWR Prairie tank, No. 4150 awaits collection from the stabling point sited in the old goods yard in Barry docks.

At least four different eras become apparent on the fading paintwork of this 'S15' tender. Two styles of 'Southern' lettering are visible and the two British Railway's lion and wheel emblems start to appear.

of the period, sent prices of locomotives sky high; some of the larger locomotives were selling at £12,500 by 1980.

I was able to visit Dai Woodham at his house shortly before his death in 1994. He was in a buoyant but reflective mood and said how he found the whole saga quite fascinating. Although he did say that if he knew in 1968 the pressures he would endure for the next twenty years, he would never have contemplated getting involved with selling the locomotives. He had great respect for those coming into his office, pleading with him to reserve a particular locomotive whilst they raised the purchase price and for the shear dedication of the men and women who came to his yard to work on a locomotive. An occasional visit to see one of 'his' engines working again on a preserved line, quite clearly gave him joy.

Barry town will soon echo to the sound of locomotives once more as the Vale of Glamorgan Railway Co. starts running steam trains from Barry Island. A fitting tribute to Woodhams are the ten ex-Barry locomotives stored ready for restoration and eventual use, and a new station, 'Woodham Halt'. Sadly Dai has gone, the family scrap business has gone, but the legend of Barry will live on forever.

Roger Hardingham

This photograph sets the scene at Barry, South Glamorgan. We see the upper and lower yards of Woodhams Bros. during the summer of 1972. The upper yard, formed of three inclined sidings in the foreground, was a storage area for locomotives arriving from various parts of the Southern and Western regions from 1964 onwards. These sidings were just to the east of the old Barry Works site. The lower yard, in the distance, offered more space for the influx of redundant locomotives and was known as the West Pond site. This was reclaimed land, infilled in the 1950s, but the ten or more sidings laid for dock use, were eventually leased to Woodhams in 1961. In the late-1990s the West Pond site has been completely redeveloped for housing and leisure facilities.

An early photograph of the upper yard at Barry taken on 20th September 1964, with remains of the Barry Works on the left. This yard had only just been commandeered for storage use, as space was running out in the West Pond site. Around 100 locomotives were now on site at Barry and from 1964 many from regions other than the Western were beginning to arrive. This Collett 2-6-2T, No. 4141, arrived the previous December and would spend 10 years at Barry before departing for the Severn Valley Railway along with Nos. 4930, 5164 and 7819. Such was their condition, they were allowed to travel by rail.

Photo: Barry J. Eagles

Sidings adjacent to the lower yard looking towards the east. A Churchward '4200' Class tank is at the head of a siding full of other ex-GWR tanks in September 1964. Four of these 2-8-0Ts went to Woodhams at Barry and two others were scrapped in 1965 at R S Tyley's yard close to Woodhams' upper site. It was during this period that numbers of Western engines were being incorrectly reported. This one was thought by many to be No. 4278, but later research confirmed that this engine never went to Barry. Only Nos. 4247, 4248, 4270 and 4277 were purchased by Woodhams. *Photo: Barry J. Eagles*

In June 1964 11 ex-Southern locomotives arrived in this Western territory. They comprised six 'S15s', four 'Us' and one 'N'. This Maunsell 'S15', No. 30841, had been withdrawn from Feltham shed in the January of 1964 and is seen just eight months later in the upper yard. Note how most of the copper and brass fittings are still in place on the engine. These would be removed soon afterwards by Woodhams, due to their value. *Photo: Barry J. Eagles*

The sidings in the upper yard, ten years later, in December 1974. From 1964, as Woodham Bros. bought locomotives from BR, they were usually stored in this yard, which had direct access to the main line from Cardiff to Barry. The town in the background is dwarfed in this view by ex-Southern S15s, many of which were withdrawn in the early part of 1964. Woodhams' offices were 100 yards to the right of this view, with an address of simply - No. 1 Dock, Barry.

Fortunately for railway historians, the Barry yards were host to some rare breeds of locomotive. Ex-Somerset & Dorset 2-8-0 No. 53808 had only been in the yard for three months when this photograph was taken in September 1964. The Fowler '7F' had been withdrawn from Bath shed in the February of 1964 after 39 years of hard work on the S&D line. Again, most of its non ferrous fittings are still attached, but it would have to wait until 1970 before rescue by the Somerset & Dorset Circle, later to become a trust. No. 53808 was initially transported back to S&D territory at Radstock. A little later, a move to Washford on the West Somerset Railway gave the Circle the opportunity to restore the locomotive back to working order.

Two locomotives are viewed on the inclined siding in the upper yard in September 1964; Nos. 4277 and 31618. Note how someone has marked up the cabside of 4277 with its number. There was no problem in identifying the 'U' Class No. 31618, which was the second locomotive to leave Barry for preservation. It left the yard by rail in January 1969 to New Hythe in Kent and spent some time at the Kent & East Sussex Railway. But it is more associated these days with the Bluebell Railway, where it reached in 1977. No. 4277 was purchased privately and moved on to the Gloucester, Warwickshire Railway in June 1986, relocating to a private site for final restoration.   *Photo: Barry J. Eagles*

The western-most point of the West Pond site in 1970. This was by far the largest of the storage areas used by Woodhams to store their locomotives 'for a rainy day'. Locomotives in view are Nos. 46428, 80151 and 71000 *Duke of Gloucester*. The general appearance of the engines in the yard at this time was fair, compared to the poor condition they were in during the 1980s. Only 12 locomotives had been bought for preservation and moved away from Barry by 1970. No. 46428 had arrived at the yard in September 1967 and would wait until October 1979 before departing for the Strathspey Railway. No. 80151 also arrived at Barry in September 1967 (from the Southern Region) and departed for the Stour Valley Railway in March 1975.

SR Class 'N' No. 31874 in the upper yard in 1973. This sole surviving member of this 66-strong class, bears a recent legend on its tender painted on for its new owner, John Bunch of the Mid-Hants Railway. No. 31874 was the 48th locomotive to leave the yard for preservation, arriving at Alresford in March 1974. After a very short overhaul period of two and a half years, it became the locomotive used to haul the Mid-Hants Railway's re-opening train on 30th April 1977.

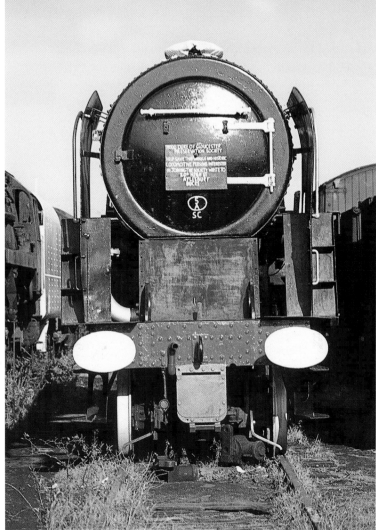

Left: From the tender of Urie 'S15' No. 30506 in the upper yard. By now (1973) all the non Ferrous fittings had been removed and scrapped. Note the rods for 30506 are still in the tender, saving the cost of replacements. No. 30506 had the unfortunate honour of hauling Nos. 30499, 30841 and 30847 to Barry from Feltham shed in June1964. However, as it got to Staines it failed with a blown superheater element and had to be towed to Barry itself later that month.

Right: No. 71000 *Duke of Gloucester* pictured in 1973. After just eight years in BR service this historic locomotive arrived at Barry in October 1967. It was in fact withdrawn by BR in 1962 and stored at Crewe. Not of course realising that in the years to follow this and other locomotives would be the subject of preservation efforts, the authorities allowed the Science Museum to remove its two outer cylinders for display purposes. The preservation of this unique locomotive was always considered 'mission impossible' in those early years at Barry. But everyone was proved wrong when the owners completely overhauled the locomotive to main line condition, with brand new cylinders.

Two ex-GWR Class '2884' 2-8-0s reside in the West Pond sidings in 1970. The identification of GWR types was extremely difficult for those roaming the seemingly endless lines of locomotives. Without their smokebox and cabside numberplates, it was virtually impossible to tell which one was which; only the numbers stamped on the various coupling and connecting rods gave some clue as to their identification. This 2-8-0 has No. 3855 chalked on its cabside after several attempts at trying to identify it. As the Barry phenomenon grew in popularity, enthusiasts started to research into their individual histories in more detail, with records of sales at the various BR works and depots being consulted. One group, the Urie Locomotive Society, began to publish a booklet 'The Barry List' in 1973 detailing the numbers and basic details of every locomotive in the yard. Indeed, until this booklet became available, Woodhams Bros. themselves didn't really have a full account of the stock on their books!

LMS types arrived at Barry from 1964 until the end of steam in 1968. This one, Stanier '8F' No. 48305, arrived at this seaside location in September 1968 and like all those arriving during this period, went straight into the larger lower yard, as by now the upper yard was full of condemned locomotives. No. 48305 remained here for 17 years before finding a new home at the Great Central Railway. Before it was purchased it became famous for an emotional inscription on its smokebox door - 'Please Don't Let Me Die'. It operated on the LMS and British Railways for 25 years and happily has enjoyed further working periods on the GCR following a full restoration.

Six War Department tanks were bought by Woodham Bros. in 1963. This line-up in the West Pond site, includes the three from the Longmoor Military Railway in Hampshire; Nos. WD108, 178 and 203. These 0-6-0 saddle tanks actually arrived by road! They were all scrapped by Woodhams by 1965. Some 42 locomotives were scrapped at the West Pond site in 1965 and approximately 40 'new' locomotives arrived from BR sheds, giving an accumulation of around 150 locomotives in both yards at Barry. The year 1965 was to see the end of wholesale scrapping of engines at Woodhams which signalled the gaining of lucrative wagon contracts, something which would keep the scrapmen at the yard busy for over 20 years. *Photo: Barry J. Eagles*

The real atmosphere of a scrapyard is pictured here in this scene from July 1975. Nearly 70 locomotives had by now departed from the yard for pastures new which caused considerable upheaval within the two sites operated by Woodhams. Both yards were by this time beginning to empty as engines departed and to achieve this engines had to be shunted to and fro to get a particular one out of the line for its transport away. Some locomotives, such as Maunsell 'S15' No. 30847, found themselves in the lower yard after many years languishing in the upper one. This locomotive was the very last 4-6-0 built under the Southern Railway flag and, like many of the 'S15s' at Barry, was withdrawn from Feltham in January 1964. It travelled to Barry in the company of other members of the class and was purchased by the Maunsell Society and steamed once more at its new home on the Bluebell Railway in October 1992.

Hughes 'Crab' 2-6-0 No. 42765 was one of two members of its class to reside at Barry. This locomotive operated for almost 40 years prior to arrival at the yard in July 1967. Sister locomotive, No. 42859, arrived in the June of 1967. There are visible signs in this 1977 view that efforts are being made to hold off the effects of corrosion on No. 42765, which was eventually moved out of Barry in 1978 for a new life on the East Lancashire Railway.

One of four Class '4MT' Standard's to go to Barry. No. 75014 arrived at the yard in October 1967 after withdrawal in December 1966. As the legends painted on its superstructure indicate, it was destined for the North Yorkshire Moors Railway and arrived at this railway in February 1981. Restoration to full working order has since seen the locomotive being used on the main line once again, most notably in Scotland.   *Photo: Glynn Hague*

Above: December 1974 and still the yard was host to over 150 locomotives. No. 31625 had lost its tender by now, certainly it arrived with one in June 1964 when it was hauled by sister locomotive No. 31618 with classmates Nos. 31806 and 31638. This journey would be No. 31618's final duty under BR ownership. All would subsequently depart from Barry between 1969 and 1980. No. 31625 travelled to the Mid-Hants Railway in March 1980 and returned to steam and main line condition, in 1997.

Right: Bulleid meets Hughes in June 1977.

Without BR 'Standard' Class 5 No. 73129 going to Barry, no example of a Caprotti valve geared 'Standard would have survived. Derby-built No. 73129 was bought by Derby Corporation and moved out by rail in January 1973 to the city. It was subsequently moved to the nearby Midland Railway Centre in 1975 for restoration.

One of the eight 'Battle of Britain' Class Pacifics sent to Barry in the 1964/65 period. In total, 18 Light Pacifics and 10 'Merchant Navy' Pacifics went to Barry - all of them would be bought and dispersed around various preserved railways in Britain. One 'Merchant Navy', No. 35029 *Ellerman Lines*, resides in the National Railway Museum at York and will never operate again as it is in a sectioned condition. Just in view on the right is ex-GWR Pannier tank No. 4612, one of only six Collett Class 5700's out of 20 to survive in the yards at Barry for eventual preservation. The other 14 were scrapped at various sites in the Barry Dock area.

Above: Not quite the end of the line for this Bulleid Light Pacific. This is the type of view so well remembered by those visiting the 'graveyard of steam'. Dai Woodham often recounted that he estimated that over 2 million people visited the site during the occupancy of the yard.

Top and bottom right: When studied in detail, the site at Barry afforded some alternative photographic opportunities. The rustic colours of the boiler backplate of this Bulleid gave this plant a zest for life. The severed coupling rod belonging to No. 6990 *Witherslack Hall* would be of little use in the future.

The green livery of this famous inhabitant of the yard is still recognisable. 'Jubilee' Class No. 45699 *Galatea* looks in a sorry state in September 1974. It has lost its chimney and has had its centre driving wheels cut through. This unfortunate occurrence was the result of a derailment during a shunting manoeuvre in the yard whilst attempting to release other locomotives for transporting away. Rather than hire expensive re-railing equipment, thought unnecessary for a 'scrapped' locomotive, the wheels were simply cut through to allow shunting to restart. This type of accident also occurred to ex-GWR 'King' Class No. 6023 *King Edward II* (inset picture, but the number 6000 was incorrectly inscribed). Both locomotive's escaped cutting-up and have had new wheels fitted to replace the dismembered ones.

An ex-LMS '8F' and a 'Standard' Class 4, No. 76084, overlook the western end of the West Pond site. The red lead paint applied to the Standard will help prevent further corrosion in the hostile environment which was close to the open sea. No. 76084 left the yard in 1983 and was the 144th engine to do so. After many years in the private owners' back garden, it has moved to the Morpeth engineering works of Ian Storey.

Only one ex-LNER type ever reached Woodhams yard; this example survived due to the locomotive being used as a stationary boiler at Colwick after its withdrawal in 1965. No. 61264 was outshopped in December 1947 and was photographed in the lower yard about a year before it left for the Great Central Railway in 1976. Purchased by the Thompson B1 Locomotive Society, this locomotive has been at the centre of one of the most courageous preservation attempts. Faced with the almost total replacement of the boiler, at enormous cost, the Society has spent some 20 years fund-raising to get this 'B1' into working order. Who would have thought that this engine, viewed at Barry's 'departure' siding in June 1975, would now be operating on the Fort William line and other lines around the Britain.

One of the Maunsell 'S15s' after it had been moved to the lower yard. No. 30828 arrived at Barry in June 1964 and spent 22 years at the seaside resort. In an attempt to procure a locally-built locomotive for the town of Eastleigh, the Eastleigh Railway Preservation Society was set up and No. 30828 bought. After arrival at the Locomotive Works in the town, eleven years were spent restoring it to working order, rather than the static exhibit is was originally intended for. Under the watchful eye of former Erecting Shop Foreman, Harry Frith, the locomotive returned to steam in 1992 and the main line in 1993. Following the death of Harry Frith, the locomotive appropriately now carries his name and is based on the Swanage Railway.

A view of the 'cutting line' in November 1974. Some of the locomotives were beginning to spill out their asbestos boiler lagging as the casing began to deteriorate. Shortly after this time, a directive from a Government office was to make Woodhams employ professional people to clear up the threat of contamination in the yard. At the far end of this siding, several men would be heavily involved in the cutting up of wagons, which in turn kept the cutting torches away from the locomotive stock. Many wagon contracts had been won over the years by Woodhams following the rationalisation programme of British Railways announced in the mid-1950s. There were gaps in wagon contracts, however, which had a devastating effect on some locomotives, as we shall discover later on in this book.

There was a further collection of locomotives stored on a site within the old goods yard at Barry adjacent to the lower, West Pond, site. This was used to store locomotives 'in transit' ie. following purchase, but just prior to road haulage out of Barry. The locomotive nearest the camera is 'Battle of Britain' Class No. 34081 *92 Squadron*, which resembles the victim of an aid raid attack! But happily this locomotives moved out of Barry in November 1976 and was returned to working order in 1998 following a long overhaul at the Nene Valley Railway.

Above left: The green livery of this unrebuilt Bulleid Pacific has worn well over the ten or more years it had been at Barry. Eight unrebuilt Pacifics had been in the yard at its peak.

Above right: By 1981 almost all the locomotives had lost any paintwork they had arrived with and the rather colourful rust finish was taking over. No. 5553, has the distinction of being the very last locomotive to leave the yard, on 31st January 1990. It had been at Barry for nearly 30 years! It is now on a private site in the Birmingham area.

Another view of the temporary site close to the road. Ivatt tank No. 41312 is photographed in 1972 along with 'West Country' Pacific No. 34016 *Bodmin*. To the rear of this, is one of the four diesels to have been in the yard, No. 601 *Ark Royal*. This locomotive was scrapped in 1980. *Bodmin* was to depart for Quainton Road in July 1972, followed by a further move to the Mid-Hants Railway in November 1976. No. 41312 remained at Barry for a further two years, moving in August 1974 to Caerphilly. Ironically, this too moved location to the Mid-Hants Railway and has been passed for main line running in 1999.

One of the ex-GWR Churchward Class '2800', No. 2857, clearly giving the appearance of new ownership. Having already done 45 years hard labour on the GWR and Western Region of BR, a new home on the Severn Valley Railway awaited further work for this 1918-built locomotive.

One livery the Bulleid's didn't get in service was red. However, at Barry dedicated owners/enthusiasts spent hours in the South Wales port protecting their either recently purchased locomotive, or one which had been earmarked for them. Dai Woodham was very obliging when it came to letting dozens of men and women into his yards for this purpose. It is questionable whether this weather-proofing was of any use, but it was an exciting time for a new owner and one wanted to do what you could to improve the locomotive's appearance. In this photograph, 'West Country' Class No. 34010 *Sidmouth* sits in the main lower yard in June 1977. It would depart for the North Yorkshire Moors Railway five years later in November 1982. This large gap in time was quite often a case of owning groups having to raise funds to pay for the costly road haulage out of the Docks. Again, Dai Woodham put little pressure on groups who had perhaps owned their locomotive for years, only to keep it in his yard for long periods, rent free!

Quite a mixture of motive power in this view. Ex-LMS '8F' No. 48624 bears the slogan 'SOLD' on its cabside and a Bulleid Light Pacific, No. 34070 *Manston*, with many other markings on its side. Sandwiched in between these two locomotives is an ex-Southern 'C2X' tender, No. DS 70183, from Guildford, which was clearly bought by Woodhams. The subject of tenders for locomotives was problematical at times, as in the early years of locomotives arriving, Woodhams sold off some of the larger tenders to a steel works in South Wales for use as carriers. The one from No. 71000 *Duke of Gloucester* was perhaps the most notable. When this locomotive was purchased, the owners had to buy a tender from a '9F', putting the problem of finding a tender onto the buyers of that locomotive. But that was the name of the game in the 1970s and early 1980s - it was very much a case of 'first come, first served'.

Another photograph of the cutting up line in November 1974. A Collett 2-6-2T looks ominously towards the cutting area. Three other companies and the BR Works at Barry, cut up locomotives in the area from 1950 to 1980. Altogether 107 locomotives ended their days at Barry, 84 of them in Woodham Bros. yards; 12 in J O Williams Ltd., Barry Dock; 2 in I C Woodfield, Cadoxton; 4 in R S Tyley Ltd., Barry and Cadoxton and 5 in the BR Works. From the first inhabitants of Woodhams yard way back in 1959 up to the latter half of 1965, most locomotives were scrapped shortly after arrival. There were exceptions, such as No. 5552 which arrived at Barry in 1961 and survived through to preservation in 1986! There were others from 1962 which, along with No. 5552, must have been at the end of long lines of locomotives, therefore avoiding the men with their cutting torches.

Above left: Camera angles and photographic possibilities at Barry were endless. This view from August 1981 shows a Bulleid unrebuilt Pacific in the middle of the lower yard. Its thin air-smoothed casing is showing signs of major deterioration! Fortunately for its new owners, this item is by far one of the simplest jobs ahead of them in the locomotive's restoration.

Above right: A driver's eye view of a line-up in the yard. This ex-GWR 2-8-0, No. 3803, was moved out of the yard after 20 years in November 1983.

Above: Some residents of Barry had the spectacle of looking out over a yard full of locomotives for more than 20 years. By 7th August 1981 the number had reduced to 78, just over one third of the original total in 1968. Maunsell 'S15' No. 30825 on the left waits patiently for a buyer. In the event, it would give up its boiler only to the Urie Locomotive Society, with the remains going to the North Yorkshire Moors Railway. The Tender would be acquired for Urie locomotive No. 30499. The 'Standard' Class 4 on the right, No. 75079, would move out in March 1982 to the Plym Valley Railway. This locomotive was dubbed '75029' during filming at Barry for a documentary on David Shepherd's locomotives, 75029 and 92203 (neither from Barry of course).

Right: 'Merchant Navy' Class Pacific No. 35018 *British India Line* looks relatively complete here in June 1977. It was the first of its class to be rebuilt from original form and the first to arrive at Barry in 1964. Exactly 10 of the 30 'Merchant Navies' went to Barry and all have been purchased for various destinations around Britain. No. 35018 was the second 'Merchant Navy' to leave Woodhams and was delivered to the Mid-Hants Railway in March 1980.

The second of the two remaining Urie-designed 'S15s' which went to Barry. No. 30499 is seen in the lower yard in April 1976 coupled to a six-wheeled tender attached to it only a few months before it left Feltham for South Wales. Having bought No. 30506, the Urie Locomotive Society decided to buy No. 30499, thereby owning both remaining examples of Urie's original design. It made the trip to the Mid-Hants Railway in November 1983.

One of the Collett-designed '2884' Class, No. 3850, in the company of a 2-6-2T. It was one of 44 locomotives arriving at Barry in 1965, approximately half of which were ex-GWR types. Roughly the same number were cut up at Barry during 1965. Note how it still retains its right-hand side coupling rods, of great benefit to its new owners, the West Somerset Railway.

The 100th locomotive to leave the Barry yard. Ex-GWR 2-6-2T No. 4110 was purchased by the GWR Preservation Group and moved to Southall in May 1979. This was a great milestone in the history of preserving locomotives from Woodhams' scrapyard. Shortly after this a more consertive effort was put in to sell off the remaining 113 locomotives - The Barry Rescue Project was just around the corner, but so too were more scrapped engines.

Above left: Precisely 99 ex-GWR type locomotives survived either into preservation, or in the yard, into the 1980s; 17 Collett and Hawksworth 'Halls' were among them. No. 6984 *Owsden Hall* was one of 15 locomotives to depart from Barry in 1986, going to a private site at Bicester, Oxfordshire. By August 1981, the date of this photograph, many engines had prospective buyer's names, telephone numbers and even addresses painted on them!

Above right: Just three Fowler '4F' engines were among the 35 LMS types at Barry. Sister locomotive, No. 43924, had the distinction of being the very first locomotive to leave the site at Barry for preserved status in September 1968, barely a month after steam traction ended on BR. Indeed, some Stanier types were still arriving at this time! This '4F' No. 44123, arrived at Barry 40 years after entering service with the LMS. It left the yard in December 1981 for the Mid-Hants Railway, but was relocated to the Avon Valley Railway at Bitton.

Yet another case of 'thank goodness for Woodhams'. Here, Southern 'Q' Class No. 30541 is partly painted ready for its departure from the yard. Initially it went to Ashchurch, Gloucestershire in April 1974 and moved again to its current home on the Bluebell Railway four years later. It is the only surviving member of the 20-strong class designed by R E L Maunsell. Like many ex-Barry locomotives it has been steamed in preservation, but awaits a further overhaul for another term of use.

This rooftop-type scene is one which many people will remember from the Barry era. The majority of the locomotives in view are Great Western types with a Bulleid on the far right-hand side. All of these would make it into preservation after a spell, in some cases, of over 20 years in the sea air at Barry.

Another view typical of Woodhams' yard. A variety of types are seen here in August 1981, with a '9F' on the left and a lucky Standard Class '4' tank on the right which is clearly under new management. On the 7th August 1981 there were just 78 locomotives on the site, with 11 tenders unattached. However, to balance this, some 35 locomotives had no tender attached to them, including 13 of the 14 Bulleid's present.

One of the eight GWR 'Manor' Class to arrive at Barry, No. 7821 *Ditcheat Manor*. No. 7821 was bought originally for the Gloucester & Warwickshire Railway with No. 7828 *Odney Manor* and both were moved out to Toddington in June 1981. *Ditcheat Manor* was subsequently moved to the Llangollen Railway and *Odney Manor* to the East Lancashire Railway. The first 'Manor' to leave Barry was No. 7827 *Lydham Manor* which left in June 1970, being the 5th locomotive to depart. It has since enjoyed many years working on the picturesque Paignton - Kingswear line.   *Photo Glynn Hague*

Above: One of only two ex-GWR 'King' Class locomotives to get as far as Woodham's scrapyard. It is hard to believe that No. 6024 *King Edward I* now graces the main line again following its nine years in the yard. One of 11 locomotives to leave Barry in the first three months of 1973, No. 6024 found a new home at the Quainton Railway in Buckinghamshire. Its arrival at Barry, following withdrawal by BR, was not without incident. Both of Barry's Kings were due to go to Ward's of Briton Ferry, but due to the banning of this class west of Cardiff the sale was cancelled and special permission given to allow Woodhams to buy the two famous locomotives. A short journey followed into unchartered territory towards the west from Cardiff to Barry Docks in December 1962.

Right: It's hard to believe that this ex-LMS Class '5' No. 45337 is now to be seen at work all over Britain. In this 1981 view it looks very neglected and highlights the tremendous effort put in by owners of ex-Barry locomotives. This Black Five arrived at Barry along with four others in January 1966 and was a result of Dai Woodham buying locomotives from regions other than the Western from 1964 onwards to the end of steam in 1968. *Photo: Glynn Hague*

Above: Fourteen of the popular Standard Class 4MT tanks went to Barry between 1965 and 1967. All were preserved. No. 80098 departed from the yard in November 1984 for the Midland Railway Centre and has been restored in 1999 for main line use. These locomotives were seen as ideal for preserved railway use and so it is not surprising that they were all snapped up. *Photo: Glynn Hague*

Right: No. 6990 *Witherslack Hall* was from the Hawksworth '6959' Class of the late 1940s. Indeed, it was outshopped in April 1948 and so was not really a Great Western engine in the true sense. No 'Halls' were broken up at Barry, so all 17 made it into preservation times. No. 4983 *Albert Hall* was the first of the class to depart from Barry in October 1970 - to the Birmingham Railway Museum. This locomotive was returned to steam in 1998 and re-named *Rood Ashton Hall*, following an investigation as to its real identity. *Witherslack Hall* left for the Great Central Railway at Loughborough in November 1975, where it has been restored to working order.

The departure line in April 1976. On a gloriously sunny day, the first of many that year, Urie 'S15' No. 30506 is about to be loaded onto a low loader operated by Wynn's from nearby Newport. This was a relatively common sight in the 1970s and 1980s at this location in the yard. No. 30506 was one of eight that year to be repatriated; in 1975 there were 14; in 1974, 19 and in 1973, 18. The Urie Locomotive Society purchased No. 30506 in March 1973 and in the intervening years raised monies to pay for the haulage costs. These were some £2,500, a tidy sum in those days and not that far from the £4,000 price actually paid for the locomotive. A three-day journey by road was ahead for the Wynn's contractors, who had to take the route via Tewksbury to avoid the Severn Bridge which had a weight restriction on it at the time. Alresford on the Md-Hants Railway was the destination and in 1987 the locomotive returned to steam once more; Bob Urie, the grandson of the designer, officially rededicating No. 30506 at a ceremony in the July.

Enormous shock waves were to reverberate on the railway preservation movement in the summer of 1980, with the news that cutting up had re-started at Woodhams Bros. Without much warning, '9F' No. 92085 was to be seen in the latter stages of destruction in mid July. A scene not really witnessed in Britain since the 1960s, saw this 2-10-0 gradually reduced to a pile of scrap. Only two others had been cut up at Barry since the 1960s; No. 76080 in April 1972 and No. 3817 in March 1973 (four diesel locomotives also succumbed). The threat was in fact always there for, in the absence of redundant wagons and other rolling stock in the yard at Barry for men to continue working on, the locomotives were always the fall-back position. Dai Woodham had, on various occasions, stated that the engine's were always considered 'work in hand' and if there was no other work for his employees, they had to cut up locomotives.

The other victim of the 1980 scare was ex-GWR Prairie, No. 4156. This 2-6-2T just happened to be the next in line on the track close to the cutting up area and was also disposed of in July/August 1980. This scene was immediately broadcast throughout Britain and would dramatically enhance the efforts being made to secure ALL the locomotives left in the yard. The Barry Rescue Project, set up in February of 1981, was to get into full swing and even mention was made in Parliament by the leading light of the campaign, the late Robert Adley MP. As history now tells us, all of the locomotives were indeed saved, a great tribute to all those involved.

By 1981 the general state of the locomotives had deteriorated to such an extent that some looked beyond economic repair. But the outward appearance was not the only factor in selecting an engine for preservation; the condition of the boiler, wheels and other major components was what counted. One of the first tasks undertaken by the Barry Rescue Project, was the full inspection of the unsold locomotives at Barry. Two prominent locomotive men, George Knight and John Peck went to Barry and inspected every one during the summer of 1981 and pronounced all of them fit for purchase. This gave the Project the impetus it required to 'sell' the idea of buying a Barry engine, so much so that during that year 21 locomotives departed from the yard!

Above: One of the locomotives to escape during 1981, was 'Battle of Britain' Class No. 34067 *Tangmere*. This haulier from Northampton got the job of loading *Tangmere* onto the trailer for its road journey to the Mid-Hants Railway in January 1981. One of eight 'Battle of Britain' Pacifics to go to Barry, No. 34067, is at the time of writing close to a main line return to steam following overhaul at Bury, Lancashire.

Right: A view looking west towards the road and rail bridges in the distance linking Barry Town with Barry Island. Since this 1981 photograph, the Vale of Glamorgan Railway Company has set up business at Barry Island, so steam train operation will return to this scene.

Oliver Bulleid would turn in his grave if he saw the condition of his 'Merchant Navy' No. 35010 *Blue Star* at Barry in August 1981. Built at the height of World War Two in July 1942, *Blue Star* was one of the last 'Merchant Navy's' to arrive at Woodhams, in March 1967. It accompanied Nos. 35027 and 35029 to the yard. No. 35027 *Port Line* was rescued in 1982 and was overhauled in the old Swindon Works complex. It now resides on the Bluebell Railway. No. 35029 *Ellerman Lines* left Barry quite early on, in January 1974 and is very unlikely to steam again, as it is in a sectioned form in the National Railway Museum at York. Happily *Blue Star* is likely to operate again; it left Barry four years after this photograph, in January 1985, initially going to Woolwich. Its owners, the BR Enginemen's Locomotive Group, have since decided to move it to the Colne Valley Railway, where full restoration is underway. Their other locomotives, Class '5' Nos. 45163 and 45293, are also at this site.

The rustic colours of No. 34046 *Braunton* are enhanced by the colour of the small tree in the foreground. *Braunton* had been at this most-westerly point in the lower yard for many years. Its paint has virtually all gone, save the area on its cabside, allowing part of the number to be recorded. Its 22-year stay at Barry was to come to an end in July 1988, with removal to the Brighton Works Project. Sadly, this scheme failed and ownership passed over to the West Somerset Railway, where the locomotive is undergoing an intensive overhaul.

One of the 14 BR Standard Class '4' tanks to reside at Barry, No. 80080, towers over the adjacent scrap piles one very cold day in November 1980. This locomotive was destined for the Peak Rail Project. Bought by Brell Ewart, the BBC Childrens programme Blue Peter followed its move to Derbyshire, but in the event it was transferred to the Midland Railway Centre nearby and fully restored. No. 80080 made a triumphant return to Barry on a main line tour in 1991, Dai Woodham was invited on the footplate for part of the trip.

One little locomotive desperate to get away, was ex-GWR Class '4575' No. 5553. This engine had been at Barry since March 1962 and had the distinction of being the very last to leave the yard in January 1990. It was the 213th, and last, time that the town of Barry witnessed a locomotive departing. Its escape to the outside world took it to Birmingham.

The general decay and noticeable lack of parts on the engines is evident in this shot. No. 34028 *Eddystone* would have to languish in this area of the yard for a further five years. It was recorded as being reserved for the Southern Pacific Rescue Group in 1981 (the date of this photograph) and like many of the Bulleids, was without a tender for most of its stay at Woodhams' yard. It was purchased by the 34028 Eddystone Locomotive Group and moved to Sellindge, Kent in May 1986. ?????

Full List of Locomotives Preserved From Barry Scrapyard

Ex-GWR Types

Churchward 2-8-0
2807
2857
2859
2861
2873
2874
Collett 2884
2885
Collett 5700
3612
3738
Collett 2884
3802
3803
3814
3822
3845
3850
3855
3862
Collett 5101
4110
4115
4121
4141
4144
4150
4160
Churchward 4200
4247
4248
4253
4270
4277
Churchward 4500
4561
4566
Churchward 4575
4588

Collett 5700
4612
Collett 4900
4920
4930
4936
4942
4953
4979
4983
Collett 4073
5029
5043
5051
5080
Churchward 4200
5164
5193
5199
Collett 5205
5224
5227
5239
Churchward 4300
5322
Churchward 4575
5521
5526
5532
5538
5539
5541
5542
5552
5553
5572
Collett 5600
5619
5637
5643
5668

Collett 4900
5900
5952
5967
5972
Collett 6000
6023
6024
Collett 5600
6619
6634
6686
6695
Hawksworth 6959
6960
6984
6989
6990
Collett 4073
7027
Collett 7200
7200
7202
7229
Churchward 4300
7325
Collett 7800
7802
7812
7819
7820
7821
7822
7827
7828
Hawksworth 6959
7903
7927
Hawksworth 9400
9466

Collett 5700
9629
9681
9682

Total Locos = 98

Ex-SR Types

Urie S15
30499
30506
Maunsell Q
30541
Maunsell S15
30825
30828
30830
30841
30847
Maunsell U
31618
31625
31638
31806
Maunsell N
31874
Bulleid WC & BB
34007
34010
34016
34027
34028
34039
34046
34053
34058
34059
34067
34070
34072
34073

34081
34092
34101
34105
Merchant Navy
35005
35006
35009
35010
35011
35018
35022
35025
35027
35029

Total Locos = 41

Ex-LMS Types

Stanier 2P
41312
41313
Stanier 5FH
42765
42859
Stanier 5FS
42968
Fowler (Midland) 4F
43924
44123
44422
Stanier 5P5F
44901
45163
45293
45337
45379
45491
Stanier 5XP
45690
45699

Ivatt 2MT
46428
46447
46512
46521
Fowler 3F 'Jinty'
47279
47298
47324
47327
47357
47406
47493
Stanier 8F
48151
48173
48305
48431
48518
48624
Fowler 7F
53808
53809

Total Locos = 35

Ex-LNER Types

Thompson B1
61264

Total Locos = 1

Ex-BR Types

8P
71000
5MT
73082
73096
73129
73156

4MT
75014
75069
75078
75079
4MT
76017
76077
76079
76084
2MT
78018
78019
78022
78059
4MT
80064
80072
80078
80079
80080
80097
80098
80100
80104
80105
80135
80136
80150
80151
9F
92134
92207
92212
92214
92219
92240
92245

Total Locos = 38

Grand Total = 213

The End - 9th November 1989

A group, mainly consisting of invited guests to the last day of Woodhams' yard, watch as No. 3845 is loaded ready for its road journey to Brighton on 9th November 1989.

So, after 30 years of locomotive scrapping and 213 locomotives bought and moved out for preservation, the final day of Woodham Brothers' yard at Barry's West Pond site dawned. A small ceremony took place in a tent erected close to the cutting line on a wet and windy 9th November 1989, with some invited guests and Dai Woodham in attendance. Robert Adley MP undertook to address the assembled people and outlined the poignance of the yard and the effect that the phenomenon had on railway preservation in Britain.

The event had been timed to co-incide with the departure of the last locomotive in the yard, No. 3845 (although in fact there was one more lurking in the distant undergrowth - No. 5553 which had been at Barry for 27 years!). But the day was an important one for it marked the truely remarkable story of a town in South Wales, which had played such a crucial role in our railway history.

The town and particularly the port of Barry was central to the industrial fortunes of South Wales from the late nineteenth century,

indeed it was in 1889 that the first reclaimed dock was filled with water. And so, 100 years later, on the day of Barry's last engine departure ceremony, a fitting tribute was given to a town and one of its citizens, one Dai Woodham.

No-one would have ever believed when it all started that so many ex-British Railways locomotives would be saved for the nation by a pure stroke of luck. That luck was the constant arrival of wagons and other rolling stock at the Woodhams site; this alone kept the cutting torches away from the locomotives from 1965 onwards and gave preservationists enough time to give the yard's occupants another lease of life.

It was not quite all over for Woodham Bros. in 1989. They closed their West Pond cutting up site a few years later and finally, in 1994, their last business Woodhams Metals Ltd. Sadly Dai Woodham died in the summer of 1994, but he will be fondly remembered in the local area for his business acumen and nationally for the part he played in this great saga.

The author stands by the gate of Dai Woodhams' last family business in 1994. Roger Hardingham has had a keen interest in Barry matters since 1972. He wrote the eight editions of *The Barry List* and is founder and chairman of the Urie Locomotive Society.